THE SECOND WORLD WAR AT SEA
IN PHOTOGRAPHS

1940

THE SECOND WORLD WAR AT SEA IN PHOTOGRAPHS

1940

PHIL CARRADICE

AMBERLEY

First published 2014

Amberley Publishing
The Hill, Stroud
Gloucestershire, GL5 4EP

www.amberley-books.com

British Library Cataloguing in Publication Data.
A catalogue record for this book is available from the British Library.

ISBN 978 1 4456 2240 8 (print)
ISBN 978 1 4456 2263 7 (ebook)

Typesetting and Origination by Amberley Publishing.
Printed in Great Britain.

Contents

Introduction

As far as the men of the BEF in France were concerned, the early part of 1940 saw a continuation of the 'phoney war' that they had been enduring since the previous September. Nothing happened in the first few months of the year, just as nothing had happened in the final few months of the last, and soldiers of all the warring nations simply sat in their prepared positions, carefully watching the enemy on the other side of the lines.

At sea, however, it was a very different matter. In January 1940 alone, more than seventy British or neutral merchant ships heading towards the British Isles were lost either to German U-boats or to mines, a total tonnage of somewhere in the region of 220,000. The cruellest, the most dangerous and the longest-lasting war of all had truly begun in earnest.

On 15 February the German Kriegsmarine – in a response to the British decision to arm her merchantmen – announced that, in future, all British merchant ships would be treated as warships and, therefore, would be liable to be sunk on sight.

In order to counter the U-boat threat, the British corvette-building programme was stepped up and soon dozens of these robust little ships – escort vessels that were really nothing more than large trawlers – were rolling off the stocks in building yards all over the country. The first of the new Flower Class corvettes was launched at Middlesborough on 24 January 1940.

The Flower Class corvettes were vessels that horrified the purists. Armed with just one 4-inch pop gun and a row of depth charges, with a single mast in front of the bridge and, horror of horrors, seamen and stokers all catered for in the same mess, they may not have been the most attractive of ships, but over the next few years they were to prove amazingly effective.

The war that year may have begun slowly but it soon began to gather pace. In a long and very bloody twelve months there were several highlights or moments of note. These can be detailed as follows:

> On 16 February, in an operation carried out in the best traditions of Nelson's navy, men of the destroyer *Cossack* sailed into a Norwegian fjord and boarded the German supply ship *Altmark* to free prisoners previously taken by the *Graf Spee*.

In the First Battle of Narvik, on 10 April, a British destroyer flotilla attacked a superior German destroyer force and sank two of them as well as seven merchant ships. Three days later, the Second Battle of Narvik saw the *Warspite* and several destroyers sink a further eight enemy ships.

Following the success of the German attack on France, the end of May and the first few days of June saw the implementation of Operation Dynamo when the Royal Navy evacuated thousands of men, British and French, from the beaches of Dunkirk. The 'miracle of the little ships' was a wonderful achievement, vital for Britain's continued ability to wage war, but, no matter what gloss was put on it by Churchill and the British propaganda machine, the Dunkirk evacuation was still a defeat.

On 11 June Benito Mussolini, Il Duce of Italy, saw his opportunity and threw in his lot with Hitler. On that day Italy declared war on Britain. It meant that the huge and powerful Italian fleet had suddenly become a major threat to British ships in the Mediterranean – and to the convoy route to India.

With France out of the war, and Italy joining the Axis side, it was essential for Britain to maintain a strong naval presence in the Mediterranean. The formation of Force H under Admiral Sir James Somerville on 28 June went some way to redressing the balance and filling the gap left by the French surrender.

On several occasions during July the ships of Force H made attacks on various units of the French fleet in their North African bases. The French – now supposedly neutral – were outraged and the attacks, while removing the threat of powerful French vessels which might otherwise be brought into the war on the German side, did irreparable damage to British-French relations.

Following Italy's declaration of war, the siege of Malta began on 11 June. Lasting until 20 July 1943, the island endured 3,340 air raids and was supplied by convoys in one of the most dangerous naval operations of the whole war.

Obviously, there was more – much more – that happened in 1940, some of the events fairly minor, others of lasting significance.

The capture of various cogs, wheels and other components from a German Enigma coding machine in February was just the start – it took several more seizures before the scientists at Bletchley Park could put together a working model and break the German code. However, when that code was finally broken, it enabled naval planners and strategists to log into U-boat messages and thus save countless ships and lives.

The Norwegian Campaign, undertaken to prevent the valuable minerals and other raw materials falling into German hands, was an unmitigated disaster for British land forces. By sheer coincidence, both Britain and Germany had been planning to land troops in Norway; the Germans managed to land first and, in the short and brutal campaign that followed, poorly prepared and badly equipped British units were totally out-classed by specialist German forces.

At sea, however, it was a different matter and the German Navy lost several ships during the operations. It was a small beginning but Hitler started to lose faith in his surface fleet and, as a consequence, the U-boat building programme was stepped up.

On 10 May, the same day that the German airborne troops attacked French and Belgian frontier forts and the Panzers began to roll in the west, Neville Chamberlain resigned as British prime minister. He was ill and heartbroken by his inability to prevent war and was dead within a few months. He was replaced by the man serving as First Lord of the Admiralty, Winston Churchill.

Churchill was nothing if not robust and clear in his determination to destroy any Nazi threat to Britain. His influence was immediately felt and on the very night he assumed power thirty-six British bombers raided Monchengladbach. Churchill's intention to wage a dynamic and offensive war was obvious to everyone, not least the German High Command.

However, he had arrived in power too late to prevent the fall of France and his first real task was to oversee the evacuation or rescue of as many British troops as could possibly be brought safely back to Britain. In an operation that lasted ten days, approximately 335,000 men were rescued from the beaches around Dunkirk, at a cost to the Navy of some eighty vessels, including nine destroyers.

After Italy entered the war on 12 June, operations in the Mediterranean Sea began to assume greater significance. The need to protect Egypt and the Near East – in particular the Suez Canal, which for years had been Britain's swiftest way to India – began to dominate naval thinking.

Mussolini had long held designs on forging an Italian Empire in North Africa, an area which had for many years been the preserve of Britain and France, and the only way he could assume control of the region was by waging war.

Italy possessed a strong and powerful surface fleet, with modern capital ships that were more than capable of running the Royal Navy close in any encounter. During the 1930s she had also built up a highly effective submarine arm and it was clear, right from the start, that any combat in the 'inland sea' would be hard-fought and dangerous.

Benito Mussolini looked enviously at the success of his Fascist partner in Germany. If Hitler could so easily defeat the Allies, Mussolini thought, there was no earthly reason he could not do the same.

He regarded the Mediterranean as an Italian preserve and was eager to throw his ships against the forces of the Royal Navy. By the late summer of 1940 things were looking grim for Britain, standing alone but undaunted against the might of Fascist Germany and Italy.

January

The crew of a Bofors anti-aircraft gun. In the background is shipping unloading supplies.

Above: An enemy trawler begins to burn to water-level.

Opposite: Changing gun barrels on the upper forward turret of the battleship *Ramillies*.

3

Above: The Italian destroyer *Nicoloso da Recco*, with a German submarine alongside, is shown here lying off Tangier just before the outbreak of war. She was one of the largest of Italy's destroyers.

Opposite: A large crowd of interested sightseers lined up on the quayside at Buenos Aires to see the arrival of HMS *Achilles*, which had fought with the *Exeter* and *Ajax* against the *Admiral Graf Spee*. The photograph was taken during the ship's forty-eight-hour courtesy visit in January.

Above: The flotilla leader HMS *Exmouth*, stated by the Admiralty on 23 January to have been sunk with all hands. The *Exmouth* was commanded by Captain R. S. Benson and her complement was 175.

Opposite above: Members of the crew of HMS *Ajax*, flagship of the South Atlantic Squadron which fought against the *Admiral Graf Spee*. The photograph was taken on board the warship when she was anchored at Montevideo harbour on a two-day courtesy visit.

Opposite below: A flotilla leader of the *Greyhound* class of eight destroyers, on 21 January HMS *Grenville* was sunk by a mine or torpedo in the North Sea. Eight men were killed and seventy-three were reported missing.

The enormous fountain of water denotes that a British destroyer on patrol has launched a depth charge at a German U-boat. The men in the foreground are preparing other depth charges.

February

Looking aft from a submarine conning tower, this picture shows the aerial and jumping wire.

Above: Close up of a submarine control room. The vessel is about to submerge, and members of the crew are monitoring the process, watching one of the two diving gauges that register depth.

Opposite: Inside a British submarine as she submerges. This compartment is situated aft within the vessel.

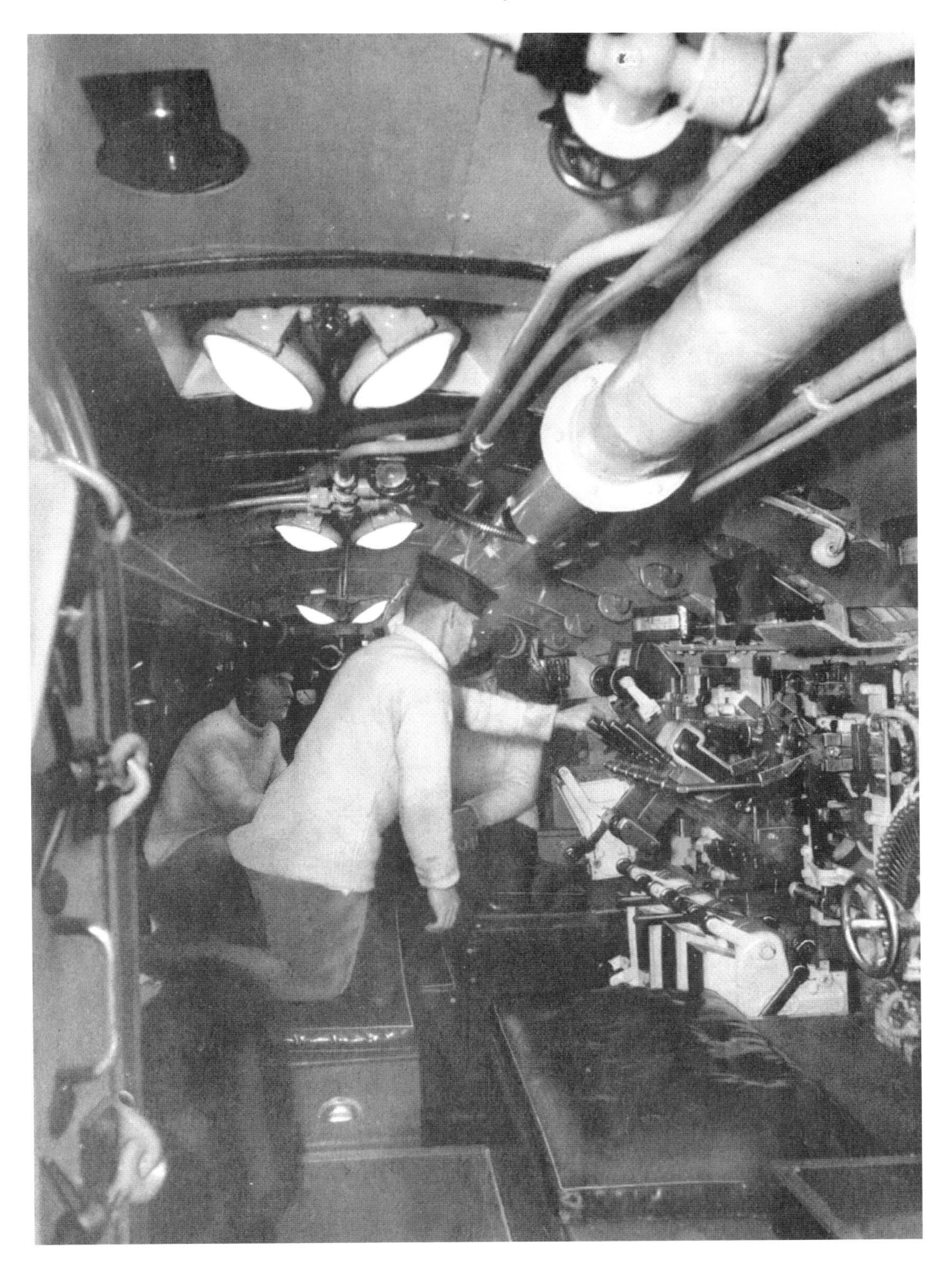

Officers receiving instruction in the assembly of a torpedo, already by 1940 one of the most deadly weapons of the war.

Physical fitness was always a golden rule in the Royal Navy. This shows bluejackets at drill during a calm and sunny afternoon.

The Polish submarine *Orzel*, which escaped from the Baltic and then torpedoed and sank the German transport *Rio de Janeiro*, which was full of troops and horses.

With its sinker in position, this shows a mine at the moment of launching, dropping over the stern of a mine-laying vessel in the North Sea. The projecting horns, which contain acid, set off a detonator when struck so that the charge explodes.

H36

Above: Sightseers watching the arrival of the *Cossack* at Leith, Scotland, on 17 February. On board the destroyer were 299 British merchantmen who had been rescued from the Nazi prison-ship *Altmark*, off the Norwegian coast.

Opposite: On 9 February the Admiralty announced that two U-boats had been destroyed by the *Antelope*. The submarines were destroyed while attaching a convoy.

This image shows the *Altmark* held fast in the ice in the narrow Joessing Fjord, a few yards from the shore. This was where she was boarded by a party of sailors from the *Cossack*. Her flag can be seen flying at half mast.

The battlecruiser *Repulse*, which the Germans claimed to have sunk on numerous occasions, arrived in Plymouth on 14 February after being at sea for 130 days. Soon after arrival, the ship's company were given well-deserved leave.

March

A Walrus amphibious aircraft has been out on patrol, scouring sky and sea. The pilot has gone below to make his report while the aircraft is hoisted on to the catapult ready for its next flight.

Above: On board a mine-layer, officers and men are making final adjustments to the mines, prior to laying them in the path of enemy vessels. An armed guard stands by – just in case.

Opposite: Lieutenant W. D. King in the conning tower of his submarine, *Snapper*, which sank several German transports and store ships while operating in Norwegian waters.

SNAPPER
39

Built in 1925, and with a tonnage of 20,277 tons, the armed merchant cruiser HMS *Carinthia* was a well-known Cunard White Star liner. When sunk by a German U-boat, two officers and two ratings were killed.

On 2 March, the largest liner in the world, the 83,673-ton *Queen Elizabeth*, left the Clyde on her maiden voyage to New York. The journey, described as a new epic of sea adventure, took five days and nine hours. The anti-magnetic mine degaussing girdle, which, when energised by electricity, neutralisd the magnetic field of the steel vessel, can be seen around the top of the liner's hull.

The hospital carrier ship *St Andrew*, which had accommodation for 300 men and was fully equipped with hospital equipment.

Painted a dull grey, with all port-holes closed, the great liner *Queen Elizabeth* (right) is seen here the moment she berthed alongside the *Queen Mary* in New York harbour.

April

Naval help for the army.

Above: A fast motor boat cuts through the waters around the coast of Eire. As the war went on, vessels of this type played an increasingly important part in sea warfare.

Opposite above: The *Gay Viking*, which ran the gauntlet of the German blockade of Sweden.

Opposite below: In this photograph, British officers in charge of a practice shoot are watching the effect of the firing. Long-range guns were used throughout the war to bombard the naval bases of the enemy.

The destroyer HMS *Punjabi*, which took part in the second battle of Narvik, berths alongside a battleship after a patrol. She is welcomed back by a band of Royal Marines.

HMS *Hostile*. In the first battle at Narvik, the *Hostile*, along with four other British destroyers, sank one Nazi destroyer, damaged three and sank six store-ships.

HMS *Hardy*, one of the five British destroyers which took part in the first battle at Narvik. The *Hardy* was so severely damaged that she had to be run ashore.

HMS *Hunter*, one of the attacking British destroyers. After a heroic battle, the *Hunter* was sunk in Narvik Fjord.

One of the most famous naval exploits of the First World War was the blocking of the Zeebrugge on St George's Day, 23 April 1918. History repeated itself in early June 1940, when – with German land forces rampaging through Europe – blockships were again sunk across the lock gates of Zeebrugge to render it useless to the Germans.

During the first attack on Narvik, HMS *Hardy*, which led the destroyer attack, was hit and had to be grounded. Undaunted, her crew clambered into the icy water and swam to the shore. This image shows some of her men helping their wounded comrades up a steep snow-covered slope.

The destroyers *Forester* and *Cossack*, seen on the left, two of the warships which took part in the second battle at Narvik. They are shown amongst the wreckage of German transports in the harbour, ships already sunk by the second destroyer flotilla.

The Second Battle of Narvik. The battleship *Warspite*, the command of Vice-Admiral Whitworth, advances up Ofot Fjord behind the six British destroyers, one of which can be seen above, on the left.

HMS *Warspite*, the British battleship which led the attack on seven enemy warships at Narvik. She was a veteran of Jutland, with a displacement of 32,000 tons and an armament including eight 15-inch guns.

Above: British troops, part of the Norwegian invasion force, are seen here on board a troopship during boat drill instruction. Tanks, lorries, heavy artillery and supplies accompanied the troops on the voyage.

Opposite: HMS *Truant*, the submarine that torpedoed and sank the 6,000- ton *Karlsruhe* on 10 April.

Men of the *Spearfish*, commanded by Lieutenant Commander J. H. Forbes (DSO), which torpedoed the *Admiral Scheer* during the Norwegian campaign.

May

The British destroyer HMS *Kelly*, commanded by Lord Louis Mountbatten, was torpedoed off the German coast in May. Badly damaged, she was towed across the North Sea to England, attacked all the way by E-boats and aircraft. By December she had been repaired and was again on active service. This picture shows members of her crew being transferred to another vessel.

One answer to Hitler's threat of invasion – a 16-inch shell is lowered onto the deck of a battleship. The shell weighed a ton and could be fired over a distance of more than 10 miles.

With her white ensign flying in the breeze, the submarine *Sturgeon*, which sank a 10,000 ton German transport off the coast of Denmark, returns to port. The German transport apparently carried between 3,000 and 4,000 troops.

Just one of the many actions of the war, the Norwegian ship *Tropic Sea* was captured by a Nazi surface raider and manned by a prize crew. She was intercepted by the submarine *Truant* off Cape Finisterre. The Nazis promptly scuttled their ship and took to the boats with their captives, among whom were the crew of the British steamer *Haxby*. The submarine took the British seamen aboard, together with the master of the *Tropic Sea* and his wife. Flying boats then picked up another boatload of Norwegians, and arrangements were made for the Germans to be taken to Spain.

Above: The New Zealand ship *Achilles* which fought side by side with the British cruisers *Exeter* and *Ajax* in the battle of the River Plate.

Opposite: Sailors cheer the submarine *Sunfish* as she returns to port after taking part in operations off the Norwegian coast. The *Sunfish*, commanded by Lieutenant Commander J. Slaughter, sank four enemy ships during her patrol.

Above: HMAS *Perth*, a light cruiser of the Royal Australian Navy.

Opposite above: At sea with the Royal Australian Navy. A seaplane is being launched by catapult during exercises. Australian naval forces were granted the title Royal Australian Navy in 1912.

Opposite below: On 4 May Allied battleships, with attendant destroyers and submarines, arrived at Alexandria, when the first part of the Allied Mediterranean fleet returned to Egyptian waters.

With the arrival of the Allied warships at Alexandria, the defences of Egypt were more fully assured. This shows an inspection of sailors and marines being carried out on board one of the British battleships soon after arrival in Alexandria Harbour.

On 17 May the 13,689-ton liner *Ville de Bruges*, formerly the American liner *President Harding*, was bombed and sunk while it was carrying refugees from Belgium. The liner is seen here on fire 12 miles from Antwerp.

June

The crew of HMS *Verity* which, during the Dunkirk evacuation, rescued nearly 20,000 soldiers.

Above: British Beaufighters attacking an enemy convoy.

Opposite: An armed trawler bursts into flame after being attacked from the air. In the background, another trawler is ringed with near-misses.

Above: The Forth Bridge spans the background to this photograph and the guns of the Royal Navy stand on guard.

Opposite above: An image taken on board the French sloop *Dubach*. Free French sailors, wearing steel helmets and general service respirators, are about to drop a depth charge in the constant war on enemy submarines.

Opposite below: Despite the French surrender, the French sloop *Dubach* served with the Royal Navy, being manned entirely by Free Frenchmen. The Free French detachments – naval and military – were eager to continue the fight with Nazi Germany.

Above: Whalers crammed with men of the BEF pulled by a motor boat to a waiting destroyer during the evacuation of Dunkirk. In the withdrawal, naval losses included six destroyers and several minesweepers.

Opposite above: A hospital ship lies sunk in a Belgian port after the vessel had been bombed and machine-gunned by planes of the German air force. According to reports, the Red Cross flag remained flying at her mast as she sank.

Opposite below: Heading for England, a tug with a motor-launch in tow, laden with soldiers from the beaches of Dunkirk. The tug and launch were just two of the hundreds of other vessels that comprised the amazing rescue armada.

Motor-launches, fishing smacks, small pleasure craft and coastal freighters taking aboard Allied troops during the evacuation. They all did heroic work in getting the men out of Dunkirk.

Above: The withdrawal of Allied forces from Narvik was covered, not without losses, by ships of the Royal Navy. Under fierce fire from enemy shore batteries and attack by aircraft, the destroyers *Ardent*, seen here, and *Acasta* were sunk.

Opposite: Destroyers packed with troops on their arrival at a British port.

One of the most modern French liners, the 28,000-ton *Champlain*, sank after striking a mine off the French coast towards the end of June. The mines had been laid only the previous night.

On 6 June the tragic action at Oran saw the battlecruiser *Dunkerque* driven ashore and later bombed by British aircraft. Churchill was adamant – if the French refused to hand over their ships to Britain they would have to be destroyed. The action soured Anglo-French relationships for years.

July

Pom-pom guns to defend against aerial attack are shown here in process of being loaded. Quick-firing weapons, the pom-poms could throw up an intense field of fire.

Above: Following the German victories in the Low Countries, most of the Dutch Navy remained in the hands of the Dutch government in exile. Here, the Rt Hon. A. V. Alexander, First Lord of the Admiralty, talks to Dutch officers and ratings at a British port.

Opposite below: The cruiser *Sumatra* of the Royal Netherlands Navy, which escaped being captured by the Germans when Holland surrendered, is shown here ready to continue operations against the enemy.

1 A German Enigma encoding machine. Engima was used to provide theoretically unbreakable communications between the U-boats at sea and their headquarters on land.

2 & 3 In the vast area of the North Atlantic, convoys could surprisingly easily slip past their pursuers and it was important not to give any unnecessary clues to the ships' position and course.

4 & 5 Equally, it was easy for sailors onboard convoys or warships to miss the tell-tale signs of a U-boat, especially a periscope peeking out of the water. Good binoculars were vital for keeping watch.

6 HMS *Cossack*, the ship that captured the *Altmark*.

7 The destroyer *Glowworm* rams the *Admiral Hipper*.

8 & 9 Recruitment posters for the Women's Royal Naval Service, otherwise known as the Wrens. Women would serve in clerical roles, and as drivers and mechanics among other things, freeing up men for combat roles.

10 An aerial view of the battlecruiser HMS *Hood*.

11 Nobody in wartime Britain was under any illusion about what the population owed to the men of the Merchant Navy.

12 The destroyer HMS *Kelly* passing the battleship *Royal Sovereign*.

13 An idealised image of the reaction of merchant seamen to the sinking of their ship by a U-boat.

14 Shelling a U-boat.

High explosive bombs bursting near a vessel during an aerial attack by German bombers in the North Sea. The escorting Allied warships successfully drove off the enemy planes.

Two French battleships of the 'Bretagne' class, 22,000 tons, were soon accounted for in a further action at Oran on 3 July. The *Lorraine*, seen here,was completed in 1916.

The Germans were quick to claim credit when the *Arandora Star* was sunk without warning on 1 July. However, the liner was transporting 1,500 German and Italian aliens, and many of them were drowned.

Completed in 1916, the French battleship *Provence* carried ten 13.4-inch guns. A sister ship to the *Lorraine*, the *Provence* also suffered in the action at Oran.

Completed in 1938, the French light cruiser *Mogador* was, with her sister ship the *Volta*, the heaviest and most powerful of her class. The *Mogador* was put out of action at Oran.

On 4 July Prime Minister Winston Churchill announced that more than 200 French vessels had been handed over to the Royal Navy and were being held in British ports. Among them, he announced, were two battleships, two light cruisers, eight destroyers and a number of submarines.

Two battleships of the 'Bretagne' class, 22,000 tons, were disabled at Oran. The *Bretagne* is shown here.

Above: Pride of the French submarine arm, the *Surcouf* was among the 200 French vessels lying in British water on 3 July. With two 8-inch guns and fourteen torpedo tubes, in 1940 she was the largest submarine afloat.

Opposite: One of the French battleships handed over to the British Navy at anchor in a British port.

A bird's-eye view of Dakar, the French West African port where the 35,000 ton French battleship *Richelieu* was attacked and put out of action by the Royal Navy on 8 July.

Above: On 14 July forty Junkers 87 dive-bombers, escorted by Messerschmitt 109 fighters, staged a concentrated attack on a convoy in the English Channel. Luckily no ships were hit. In this view two bombs are seen falling astern of a British destroyer.

Opposite below: To ensure that the recently completed French battleship *Richelieu*, stationed at Dakar in French West Africa, did not fall into the hands of the enemy, on 8 July Lieutenant Commander R. H. Bristowe was entrusted with the hazardous task of putting her out of action by destroying her steering gear. This he successfully accomplished by getting through the harbour defences in a small motor-boat and dropping depth charges. The Fleet Air Arm then made the main attack.

Units of the Australian fleet operated in British home waters. This shows Australian gunners at practice.

August

One floating fortress seen from the deck of another. A view from HMS *Nelson*, one of the most powerful capital ships in the Royal Navy.

Above and opposite: On 22 August a British convoy passing through the Straits of Dover was attacked by bomber aircraft operating from recently captured French airfields and by shell fire from the French coast. These images were taken by a photographer on board one of the escorting destroyers and show a bursting salvo, a single shell falling harmlessly between two ships.

Above: Despite Hitler's so-called 'total blockade of Britain', ships of the Merchant Navy continued to gather together in convoys. Escorted by destroyers and tiny corvettes, they brought precious cargoes of food – as well as weapons – into the country.

Above: This shows the Canadian armed merchant cruiser, *Prince Robert*, which captured the German cargo ship *Weser* off the Mexican coast. The German crew offered no resistance. The *Weser*'s cargo of coke, peat, moss and oil was Canada's first naval prize of the war and was valued at £333,000.

Opposite: A destroyer at sea cuts through the waves.

Fast German launches, nicknamed 'Stukas of the Sea', are shown here searching for enemy merchantmen and warships.

Keeping a close watch on enemy ships in Norwegian waters, despite the British withdrawal from the country, remained part of the Navy's duties. Here, a British destroyer sends a boarding party to inspect a suspicious merchant vessel.

September

Canadian officers and ratings take control of one of the old 'four stacker' destroyers handed over to the British Navy by the Americans.

American and Canadian sailors are seen here gathered together on the deck of one of the first fifty destroyers handed over to Britain by the Lease-Lend agreement.

The cruiser HMS *Arethusa*.

September 1940 and the Royal Navy patrols the 'Italian Lake'. Mussolini's claims to the Mediterranean as an Italian preserve did not prevent British battleships from continuing to sweep its waters. Triple torpedo tubes are here swung outboard in readiness for instant action.

Under an agreement announced on 3 September 1940, fifty 'over-age' destroyers of World War One vintage were transferred from the USA to Britain. In exchange, the Americans were granted leases of naval and air bases on British possessions in the Atlantic. This aerial view of Philadelphia Navy Yard shows massed formations of veteran warships ready for delivery to Britain.

Above: The tragedy of the *City of Benares*. On 23 September, eighty-three children on their way to Canada under the government evacuation scheme, as well as 211 other people, were reported lost when the liner *City of Benares* was torpedoed by a U-boat. Later, forty-six survivors were discovered in mid-ocean by a patrolling Sunderland flying boat.

Opposite below: British destroyers are seen here escorting a convoy through the Straits of Dover. A smokescreen, as well as guns and depth charges, help the convoy make it safely to port.

Despite the menace of U-boats, aerial attack and bombardment by heavy guns from French Channel ports, convoys continued to run. This photograph, taken from an escorting vessel, shows one of them heading in a rather unconcerned fashion on its way through the narrow waterway.

The first flotilla of American destroyers handed over by America to the Royal Navy entered British waters on 28 September. Some of the vessels were promptly renamed after towns in Britain and the United States.

The British crew of one of the destroyers passed on to Britain as part of the Lease-Lend Agreement are photographed here on their arrival in home waters. The transfer of crews took place in Canada and the voyage across the Atlantic, with the British sailors beginning to learn about the idiosyncrasies of their new ships, was uneventful.

Prime Minister Winston Churchill, who toured London docks on 25 September, is seen here boarding a naval auxiliary vessel.

British warships head out to sea. Their task is to bombard the Libyan ports of Bardia and Fort Capuzzo – another successful mission for the Mediterranean Fleet.

With decks awash in swirling foam, a motor torpedo boat is photographed on high-speed patrol. Fast and manoeuvrable launches like this carried out valuable work along the North Sea coast.

Above: The vessel shown here is a Greek freighter, stopped by the Canadian Navy. Her skipper greets the boarding officer with a handshake as he comes on board to inspect the cargo. Stopping and searching neutral vessels – or suspicious ones – was part and parcel of blockading the German coast.

Opposite: The British fleet makes for the open sea.

The first of a new class of six aircraft-carriers. Displacing 23,000 tons – heavier than the *Ark Royal* – the new carriers had a full complement of 1,600 men. For defence their armament included sixteen 4.5-inch dual-purpose guns.

In September 1939, the aircraft carrier *Courageous* was torpedoed and sunk. Her sister ship, HMS *Glorious*, seen here, was sunk by the battlecruisers *Scharnhorst* and *Gneisenau* while assisting in the Allied evacuation from Narvik.

Ships of the Italian Navy steaming in line during manoeuvres in Spanish waters. In 1940 Italy had the fifth most powerful navy in the world, her largest battleships being the *Littorio*, *Impero* and *Vittorio Veneto*, sister ships of 35,000 tons, completed in 1940.

Above: With an armament of eight 8-inch guns, the Italian cruiser *Zara* was completed in 1931. She carried twelve 3.9-inch, ten 37 mm and eight 13-mm guns.

Opposite above: The dangers attached to the work of mine-sweeping are vividly illustrated in this explosion of a floating mine. Hit by rifle fire from a British mine-sweeper, the mine explodes and a huge column of water gushes skywards.

Opposite below: The British battle-cruiser HMS *Renown* fires a salvo from her massive 15-inch guns.

A convoy made up of twenty-four merchant vessels, mostly colliers, is seen steaming in line in the North Sea. Lifeboats are ready for instant use, while escorting warships are busy steaming up and down the line, keeping a sharp look out for U-boats and enemy aircraft.

October

An effective answer to low-level attacks from Nazi bombers – some convoys were equipped with barrage balloons to make enemy aircraft keep a safe distance away.

This page and opposite page: On 12 October HMS *Ajax*, of River Plate fame, intercepted and sank two Italian destroyers off Sicily. She then encountered a heavy cruiser and four destroyers, one of them the modern 1,620-ton *Artigliere*. After crippling this vessel, *Ajax* was joined by HMS *York*, but the two ships lost their prey during the night. Next morning *Artigliere* was located in tow of another Italian destroyer, which promptly deserted her and fled. *Artigliere*'s crew was ordered to abandon ship before she was sunk by the *York*.

Above: Convoys were vital to the survival of the British Isles. Here, protected by the Royal Navy and the Royal Air Force, a group of Britain's merchant ships proceed through the Strait of Dover. Despite being barely twenty-five miles wide, the Channel remained in English control throughout the war – vital cargoes continued to get through.

Opposite above: The *Empress of Britain*, a 42,000-ton liner of the Canadian Pacific Line, was attacked by enemy aircraft on 26 October 150 miles off the Irish coast. The German bombers left the vessel a blazing wreck. Tugs attempted to tow her to port, but the effort was unsuccessful, and the liner was eventually torpedoed by the U 32 and sank that night. A total of 600 out of the 643 passengers on board were saved and brought to port.

Opposite below: On 28 October, Italy, alleging violations of Greek neutrality, presented a three-hour ultimatum to Greece, demanding that certain unspecified strategic points be conceded. It was a virtual declaration of war and an Italian attack was immediately launched. This image shows Greek warships shelling the Italian forces near the coast as they crossed the Albanian border into Greek territory.

A wartime launching. There is no launching platform now, no champagne christening, no cheering crowd - a new merchant ship joins the British war time fleet. Only a few workmen pause to witness the passage of the new ship into the water.

British soldiers landing in Egypt. With the widening of the war, the Middle East soon began to assume greater significance. The Royal Navy provided a strong escort for the troops.

Britain still rules the seas! A convoy of troopships reaches port in Egypt after the long voyage from England.

Above: A photograph of Admiral Sir Andrew Cunningham, DSO, the British Commander-in-Chief in the Mediterranean. Cunningham was the most successful British Admiral of the whole war.

Opposite: Despite the threat of a strong Italian navy, Britain retained control of the Mediterranean Sea – not always easily but always effectively. This photograph shows an aircraft carrier in the Mediterranean under the shadow of a battleship's big guns.

Above: The luxury liner SS *Empress of Britain* was sunk as the result of enemy action on 26 October.

Opposite above: After a concentration of enemy shipping was detected at Cherbourg by air reconnaissance, the Royal Navy shelled it heavily and a large number of fires were started in the dock area.

Opposite below: Heavy and light forces of the Royal Navy, working in cooperation with the RAF, took part in a heavy bombardment of enemy-occupied Cherbourg during the night of 10/11 October.

BO
FG
GB

Above: German E-boats, armed with automatic guns, equipped with diesel engines and having a range of 600 miles, made full use of smoke-screen tactics. This image shows E-boats at anchor.

Opposite above: Set on fire by bombs from enemy aircraft, SS *Empress of Britain* was taken in tow, but was later torpedoed and sank during salvage operations. There were 598 survivors of a total of 643 on board.

Opposite below: All of them fitted for mine laying, this flotilla of latest Italian destroyers makes an imposing array as they lie at anchor in the Italian naval port of Gaeta.

King George, keen to see for himself the country's coastal defence system, is piped aboard one of the speedy, manoeuvrable motor torpedo boats of Britain's 'Mosquito Navy'. These small boats carried two very effective anti-aircraft guns and torpedo tubes.

November

HMS *Dido* bombarding land targets in Italy.

On the night of 11/12 November, an attack was made on Italian capital ships in Taranto Harbour by torpedo-carrying planes from the aircraft carriers *Eagle* and *Illustrious* from Admiral Cunningham's Mediterranean Fleet. The attacks were decisive and immediately altered the balance of naval power in the Mediterranean.

Following the Italian attack on Greece, Britain promised military aid. Convoys carrying stores and equipment for Greek and British forces operating on the Albanian frontier were immediately dispatched. Here, in November 1940, *Ark Royal*, *Malaya*, and *Renown* are passing the Rock of Gibraltar.

On 27 November British naval forces encountered two enemy battleships, escorted by cruisers and destroyers, to the west of Sardinia. The Italian ships soon broke off the action and retired, a cruiser, two destroyers and smaller vessels being damaged before their escape. One British aircraft was lost and the cruiser *Berwick* suffered slight damage. *Above*, a British cruiser fires at the Italian ships as enemy shells fall astern. *Below*, a British destroyer and cruisers take up their positions before the beginning of the action.

The fast Italian cruiser *Bartolomeo Colleoni* was sunk by HMAS *Sydney*, north-west of Crete, in one of the most brilliant engagements of the war. The Italian ship's bows were virtually blown away.

A view of the captain's bridge on one of Admiral Cunningham's Mediterranean Fleet battleships.

A dive-bombing exercise showing a Spitfire attacking a British destroyer. The exercise is watched with interest by gunners of the warship in the foreground.

The Greek cruiser *Giorgios Averoff* was engaged in coastal defence. Though an old ship, she was hastily reconditioned for her present important duty.

Above: Two Italian battleships damaged by torpedoes of the Fleet Air Arm. The *Conti di Cavour*, shown in this image, and her sister ships were a powerful part of the Italian fleet.

Opposite above: Destroyers of the Greek navy are shown here at sea. During the early part of the Italian invasion of their country, they shelled Italian troops on the Albanian border, an operation witnessed from the heights of the island of Corfu.

Opposite below: Destroyers on duty. Much of the convoy work was carried out in wild weather. Note the enemy planes on the right of the picture.

Italy's battleship strength was reduced by half when the Fleet Air Arm attacked Taranto harbour on the night of the 11/12 November. Among the three capital ships put out of action was one of the 'Littorio' class.

The attack on Taranto, in which the Italian battle fleet was reduced to half its former strength, was a triumph for the Fleet Air Arm and its Fairy Swordfish aircraft. This image shows a Swordfish and the aircraft carrier HMS *Eagle*.

December

Greek forces occupied Santi Quaranta, the Italians' southernmost Albanian sea base, on 5 December. The same day the Italians evacuated Argyrokastro, which was taken by the Greeks advancing through Delvino, on 8 December. Many prisoners and equipment were captured. This shows a view of the town and harbour of Santi Quaranta.

A new gun ready for a British battleship. 'Proofing' tests were essential if there were not to be accidents during active service. A huge charge of cordite was invariably used – far stronger than any normal cordite charges. Workmen are shown taking cover in reinforced shelters. The complete test cost somewhere in the region of £500.

The Royal Navy on watch. One of Britain's capital ships, with its massive 15-inch guns.

HMS *Sturgeon* (640 tons) was launched at Chatham in 1932. During the war she was commanded by Lieutenant G. D. A. Gregory DSO, photographed here with his crew. Lieutenant Gregory stands in the back row, fifth from the left.

Towing a new British cruiser into the fitting-out basin, a shipyard scene endorsing the prime minister's announcement of the Royal Navy's ever-growing strength. A group of workers survey the operation during their lunch time break.

Blackburn Skua dive-bombers assembled on the deck of a British aircraft carrier at sea in readiness to take off for an attack.

Acknowledgements

Thanks go to Campbell McCutcheon and John Christopher of Amberley Publishing for the use of their extensive image collections and their expertise on the war. The images are sourced from:

Hutchinson's Pictorial History of the War 20 December 1939 – 9 April 1940
Hutchinson's Pictorial History of the War 10 April – 6 August 1940
Hutchinson's Pictorial History of the War 7 August – 1 October 1940
The War Illustrated
The Second Year of War in Pictures

About the Author

Phil Carradice was born at Pembroke Dock and educated at Cardiff College of Education and University College, Cardiff. He has worked as a teacher and a social worker and was the headteacher of Headlands Special School in Penarth.

Phil is a poet, author of short stories, editor, broadcaster and historian. He has had over forty books published on a wide variety of historical topics. His published books include *The First World War in the Air* and *A Pembrokeshire Childhood*. He is a regular broadcaster on BBC Radio 3 and 4 and on TV programmes such as *The One Show*. He presents *The Past Master* on BBC Radio Wales. He lives in the Vale of Glamorgan.